Adding Custom Actions to OxygenXML Frameworks

A TC Dojo Book

ELIZABETH FRALEY

Adding Custom Actions to OxygenXML Frameworks

SECOND EDITION

Published by: Single-Sourcing Solutions

PO Box 62122, Sunnyvale, CA 94089.

TC Dojo and name and logos are the property of Single-Sourcing Solutions, Inc.

OxygenXML is the property of SyncRO Soft SRL, oxygenxml.com

Arbortext product names and logos are the property of PTC Inc., ptc.com

info@single-sourcing.com

www.single-sourcing.com

twitter.com/SingleSourcing

First edition: August 2016

ISBN: 978-1-7326766-3-3

This book was written in DITA using Arbortext Editor and published for print, PDF, ePub, and MOBI using Arbortext Styler.

TC Dojo Books

Adding Custom Actions to OxygenXML Frameworks

Arbortext Monster Garage Books

Arbortext for Authoring: An Author's Guide to Getting Started with Arbortext Editor

Arbortext 101: Best Practices for Configuring, Authoring, Styling, and Publishing with Arbortext

Arbortext 102: Best Practices for Creating Arbortext Styler Stylesheets

To anyone who was ever told by a consultant that a plugin
was the only option.

Contents

Chapter 1.   About this book

Topics Covered in this Chapter

- Applicable software versions
- About the author

The toughest thing about learning any new software package is making the shift from what you already know to how this new software replaces processes and changes your world. It's shifting from how you think it should fit into your world to how your world will adapt to it so you can use it most effectively.

Knowledge is tribal. At Single-Sourcing Solutions, we have all benefitted from the knowledge of those around us. We started as customers for all the tools we use. We have always done everything we can to share our knowledge with the world around us.

Although I've had at least one license of OxygenXML since 2000, only 2 years after SyncRO Soft SRL launched the product, I've only recently had the opportunity to dig deep into the inner workings. When we were first starting this project, we struggled with the basics but came to deeply appreciate the way that the team at SyncRO Soft has approached the need for customers to be able to change their environment without handcuffing them to expensive custom development costs.

As a result, once we had figured out some of the basics, we decided to organize our crib notes and share them with other developers who are also fans of the product but who haven't had the same opportunity to dig deep.

If you haven't already investigated the OxygenXML Framework and the custom actions available in Author mode, you should. In this step-by-step guide, we will give you the keys you need in order to understand the basics. In very short order, you will be on your way to doing just about anything you want with your OxygenXML authoring environment.

Applicable software versions

This book is current as of OxygenXML 18.0 (released 09 August 2016).

Our examples assume a local installation, but what we describe here is the foundation for any advanced deployment.

About the author

Elizabeth Fraley is a serial entrepreneur. She's founded two companies, sits on the boards of three non-profits, and is constantly coming up with new ways to share knowledge in the technical communications and content industries.

She works as a Single-Source/XML Architect/Programmer and Mentor. She has worked in industries ranging from high-tech to government, at companies of all different sizes (from startups to huge enterprises). She advocates approaches that directly improve organizational efficiency, productivity, and interoperability. She takes an apprentice to journeyman approach with her customers, holding their hands and enabling them to stand on their own two feet. She knows that her customers want to do it themselves and that they just don't want to do it alone.

She constantly looks for ways to make tribal knowledge...not so tribal. She presents regularly at international, national, and regional conferences related to content, information architecture, and technical communication. Her appearances and publications are consistently well received by her peers. She is in the process of launching a new program designed to assist technical communicators grow their professional business and political skills because those skills are as important as technical prowess for success.

She has created two community-driven webinar series (the Arbortext Monster Garage and the TC Dojo), three YouTube Channels, and an Arbortext On Demand Vimeo channel. She has an Arbortext Community Voices podcast, and runs the Arbortext Meetup Group, LinkedIn Group, and Facebook group. She writes blog posts and publishes papers that enable the people in her communities. She developed the first mastermind series for technical communicators and, as president of Single-Sourcing Solutions, saved an internationally known Arbortext code archive from deletion. She's the founder of TC Camp, is the only unconference focused on the technical writing community and the people who support them.

She was on G. Ken Holman's overflow list for decades and has had an oXygenXML license since 2001.

Chapter 2.   What you need to know

Topics Covered in this Chapter

♦ Custom actions
♦ Techniques for using custom actions

The most popular way to approach customizing OxygenXML is through a plugin. Plugins give you a lot of flexibility and facilitates the development of extensive, complex integrations. Many vendors of content management systems have deployed plugins to ease the interaction between OxygenXML and their application.

But there's another way, a simpler way.

For most of the changes that most authors need in their environment, you can use the custom action mechanism that is part of the OxygenXML framework.

The OxygenXML framework provides a way to encapsulate changes to the authoring environment. The approach makes those kinds of changes easy to replicate and share.

For example, say you wanted to duplicate all the paragraphs in your document and change the language attribute on the new paragraphs to "ES" (Spanish) in preparation for translation. By adding a custom action to your framework, you can make a new menu item or toolbar button that does it automatically. This is something easily achievable with a short XSLT script and just as easily hooked into the document framework by using a custom action.

And that's only one of the kinds of things you can do through the custom action mechanism.

Custom actions

Custom actions are defined operations that can be triggered depending on the current context. You can further limit custom actions by defining an XPath expression that will activate the operation.

OxygenXML provides a large number of actions out of the box as well as a several open-ended actions that give you the ability to do almost anything you need to do. You can also implement your own.

To create a custom action, scroll through the list of available operations. Choose the one that is closest to the action you are trying to implement.

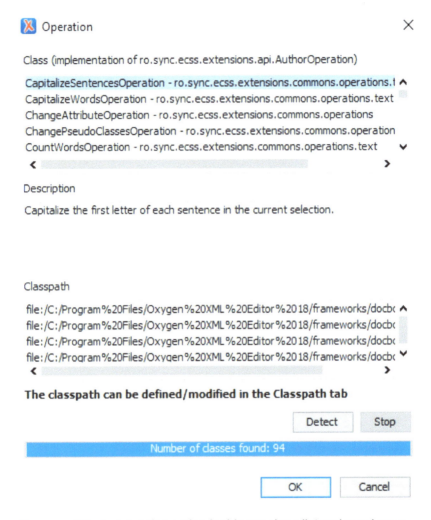

If you can't find an existing action in this very long list and need more flexibility, OxygenXML provides four actions that give you a way to choose how you want to define the operation logic:

- XSLT operation
- XQuery operation
- XQuery update operation
- JavaScript operation

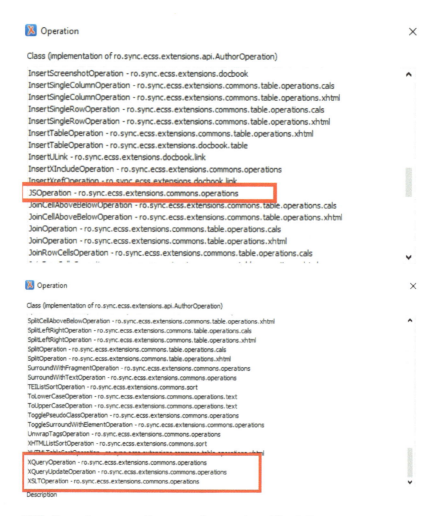

With these four operations, you have a lot of flexibility to work in a language you're already comfortable with and perform actions on the document open in the user interface. You can write a script and call it with one of these actions.

And if those aren't enough, there are a couple of other operations to take note of:

- Execute multiple actions operation
- Execute transformation scenarios operation

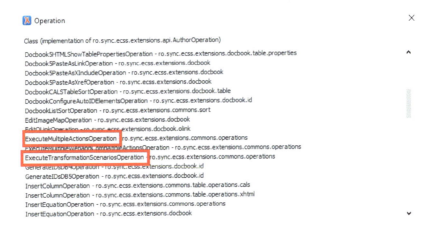

The execute multiple operations action allows you to call any number of existing actions in sequence. This includes all existing actions and any that you define. If you have something complex, you can try the multiple actions operation to split up the tasks and then knit them together. You can call all of the actions you want, in the order that you want them to happen, with the execute multiple actions operation.

The execute transformation scenario provides a way for you to define a custom transformation scenario and have it easily accessible. You can define everything about it by using this type of operation and save it as a custom action.

Between the actions that are pre-defined out of the box and the actions that give you the ability to develop your own scripts and sequences, you have a lot of flexibility and ways to get what you want.

The custom action mechanism is a powerful tool in the OxygenXML developer's toolbox. Before you go down the complex plugin path, check to see if the custom actions provide the access to accomplish what you need to do.

Techniques for using custom actions

There are a few techniques that we learned when digging into custom actions.

1. Because you can use both XSLT and XQuery scripts and because OxygenXML provides UI modes for programming with each language, you can develop and test your scripts in the IDE before you create the custom action.

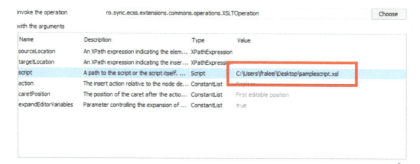

2. Pointing to a script isn't the only way to use XSLT or XQuery in a custom action. You can put the code directly into the custom action textbox. In some cases it matters which way you do it – as an external file or as code in the textbox. It's important to realize, though, that both options are available to you.

3. You can limit a custom action by defining the XPath pattern that describes the conditions under which the custom action is valid. This is a deceptively simple idea that can add finesse to an otherwise complex situation.

4. You can define multiple XPath tests for your custom action.

   The **Operations** pane of the **Action** has numbered tabs. By default, only one tab ("**1**") is created for any action. The **Operations** pane is where you can specify the XPath expression that defines when an action can be executed. This XPath expression is the test which determines whether this operation should take place.

   If you want to define another test for this action, use the **+** button in the lower right hand corner add one. Each time you click the **+**, you will add a new numbered tab to the **Operations** pane.

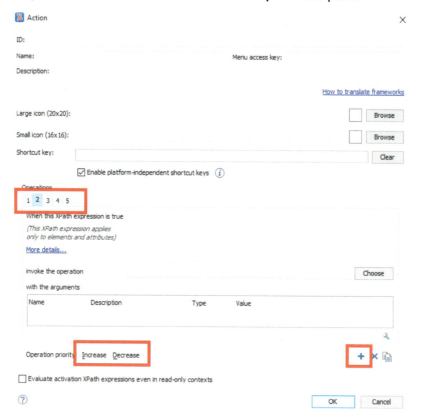

Adding multiple tests creates this structure:

```
if (XPath1) then Operation #1
else if (XPath2) then Operation #2
...
else if (XPathN) then Operation #N
```

The ordering of XPath tests is important: They define the evaluation priority. #1 happens first. If you want to change the order, click **Increase** or **Decrease** to change the order of evaluation of XPath expressions.

These techniques can be useful to remember when you're planning your approach. You don't have to do it all in one place.

Chapter 3.   Let's Do It!

Topics Covered in this Chapter

♦ Create a new custom action
♦ Add your custom action to a menu
♦ Add your custom action to a toolbar
♦ Execute your custom action

I'm going to work through a complete example of setting up a custom action. All the steps and all the code are provided for you to use.

Open OxygenXML, and let's get started.

Create a new custom action

Access the preferences window from the **Options** menu.

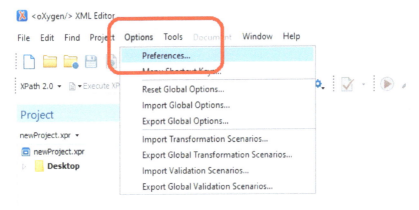

Choose **Document Type Association** in the left hand navigation and then click **Docbook 5** in the right hand pane. Then click **Extend**.

Adding Custom Actions to OxygenXML Frameworks

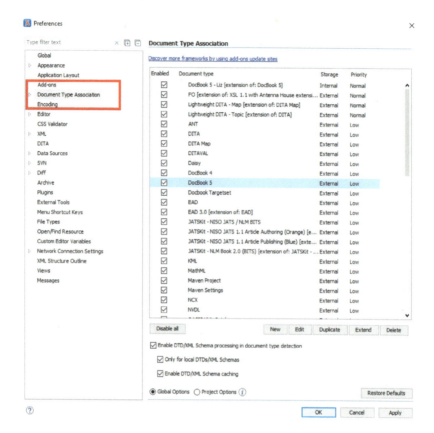

In the **Document Type** window, name your extension a name, provide a description, and click **OK**.

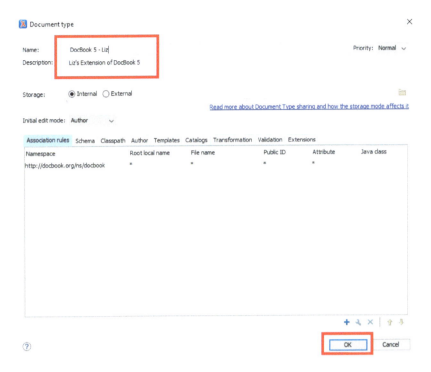

You will return to the **Preferences** window and will see your extension in the list of document type associations. Select your extension and click **Edit**.

Adding Custom Actions to OxygenXML Frameworks

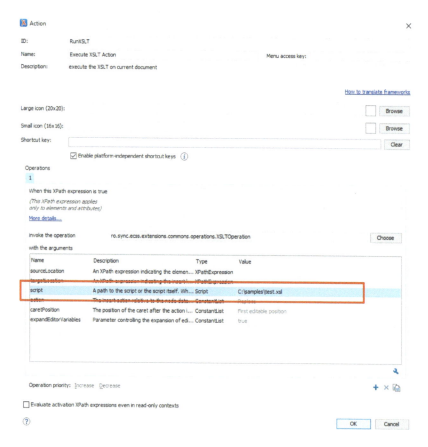

In the lower half of the **Document Type** window, select the **Author** tab and then **Actions** in the left hand column to display the list of currently available custom actions.

Any action shown in this list can be used in the execute multiple actions operation, added as a menu item or a toolbar, or accessed through a shortcut key combination.

Create a new custom action

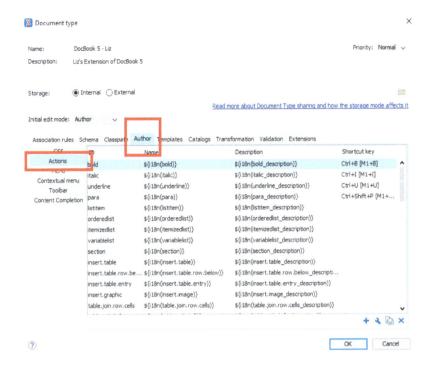

To add a new custom action, click the **+** in the bottom right corner.

Adding Custom Actions to OxygenXML Frameworks

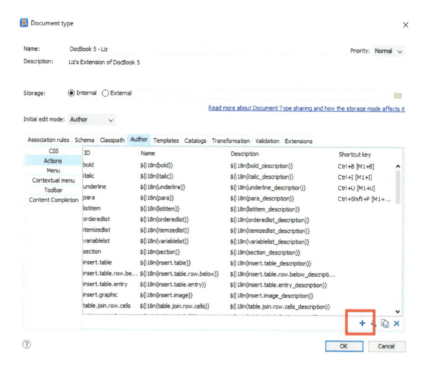

This will open the **Action** window where you define everything about the custom action.

You must provide an **ID** and **name**. You may also provide a **description** for your custom action.

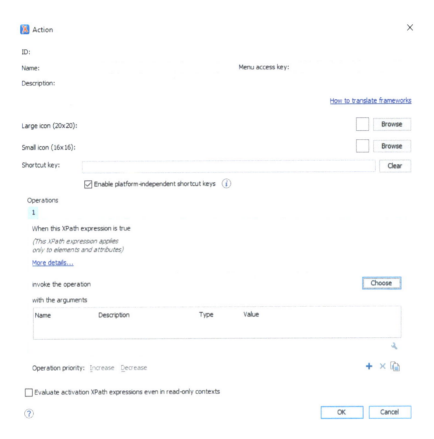

Choose the type of operation you want to invoke with this custom action.

In the **Operations** panel, click **Choose**.

Adding Custom Actions to OxygenXML Frameworks

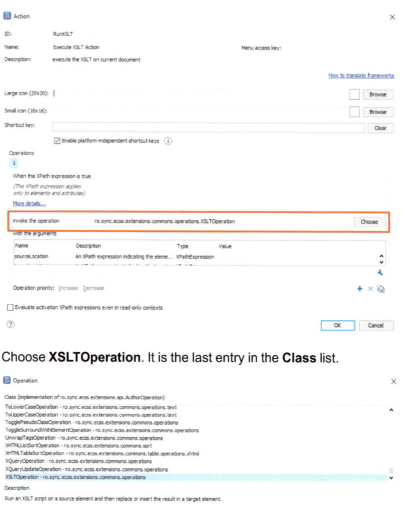

Choose **XSLTOperation**. It is the last entry in the **Class** list.

Click **OK** to close the Operation window. You will see that your custom action now appears in the list actions in the **Document Type** window.

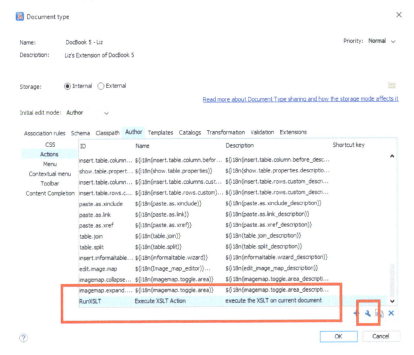

Click the **wrench** in the lower right corner to edit the action and set the arguments necessary to execute the action.

For an XSLT operation, you must define the XSLT script to be used when the action is executed. Click the **script** argument to select the argument. Then click the **wrench** to edit the argument.

Adding Custom Actions to OxygenXML Frameworks

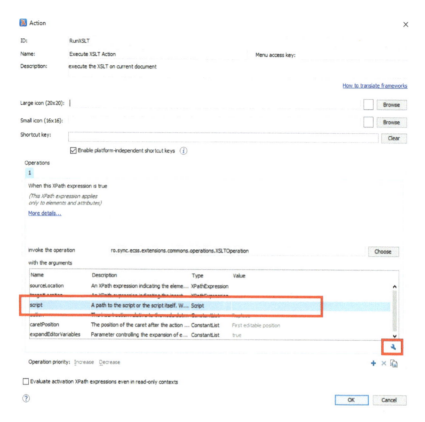

In the **Edit Argument Value** window, type the path to your XSLT file in the text box and click **OK**.

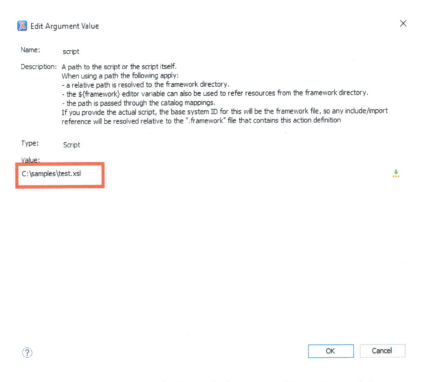

Once you are back in the **Action** window, you will see the script argument now has a value: The path to your XSLT file.

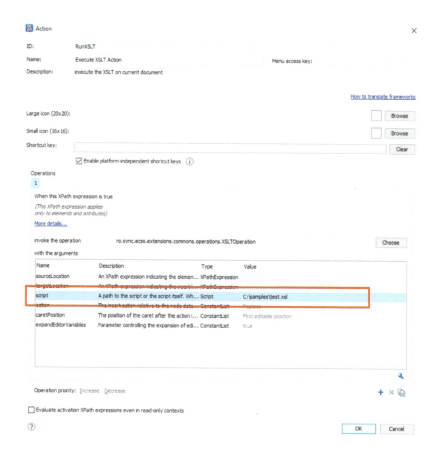

Click **OK** to save the custom action and return to the **Document Type** window.

Expert tip!

You can specify the exact path to your script, but it is generally a best practice to use a portable path that is defined relative to the framework. Although not the subject of this ebook, there are best practices for creating, extending, and storing frameworks as well. For example, by marking a framework as an **external framework**, all references to resources will be resolved relative to the framework's location.

Add your custom action to a menu

In the **Document Type** window, click the **Author** tab and then **Menu** in left hand panel.

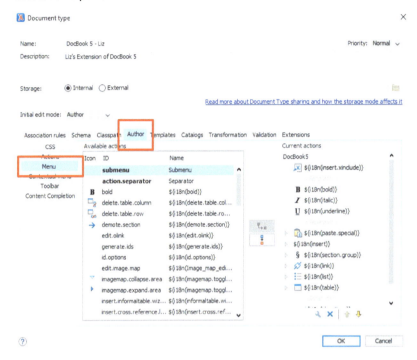

Scroll to bottom of the **Available actions** and click on your custom action.

Adding Custom Actions to OxygenXML Frameworks

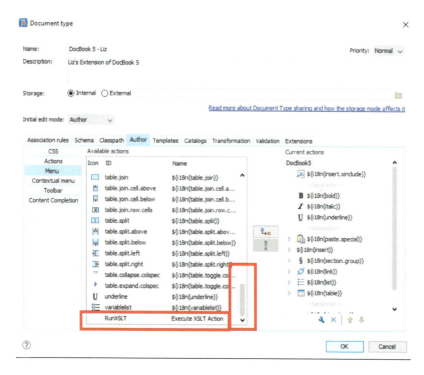

Click **Add as child**. It is the top button in between the **Available actions** and **Current actions** panel. This button will add the selected action to the menu shown in the current actions panel.

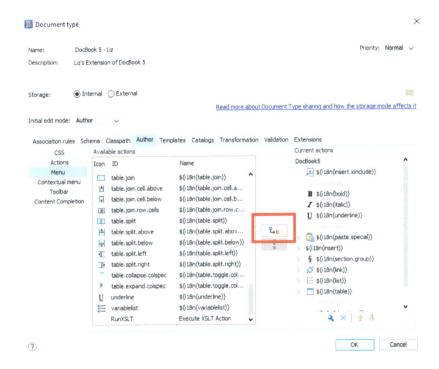

If you don't already have a custom menu, it will ask you to create one. In the window that pops up, give your custom menu a name.

Click **OK** to save the menu and return to the **Document Type** window.

You will see the new menu in the **Current actions** panel.

Adding Custom Actions to OxygenXML Frameworks

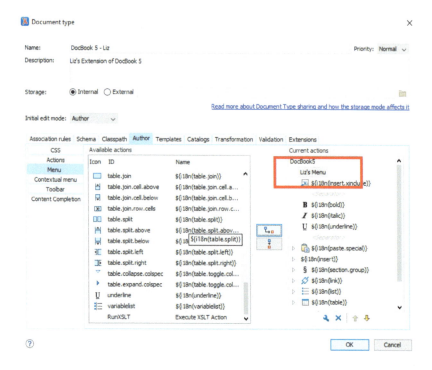

Select your action again and then click **Add as child**.

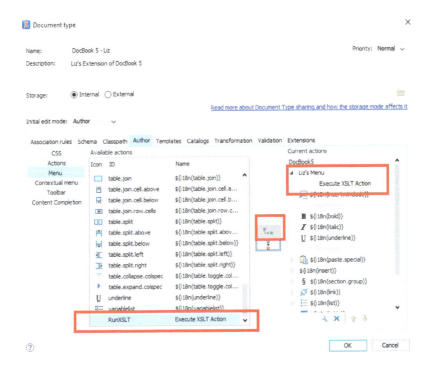

Your custom action now appears in your custom menu. Click **OK**.

Add your custom action to a toolbar

In the **Document Type** window, click the **Author** tab, then **Toolbar** in left-hand panel.

Adding Custom Actions to OxygenXML Frameworks

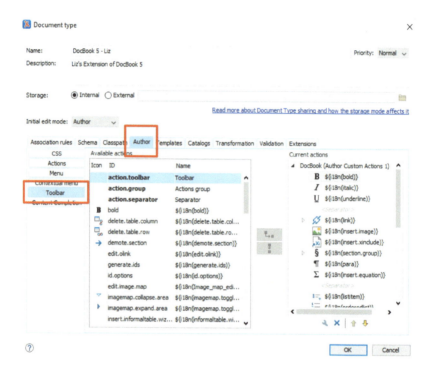

Scroll to bottom of the **Available Actions** and click on the action you want to add to the toolbar.

Add your custom action to a toolbar

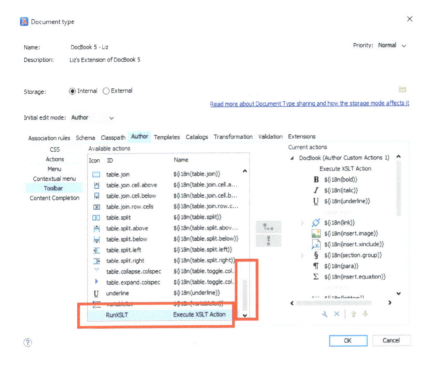

Select your custom action and click **Add as child**.

Adding Custom Actions to OxygenXML Frameworks

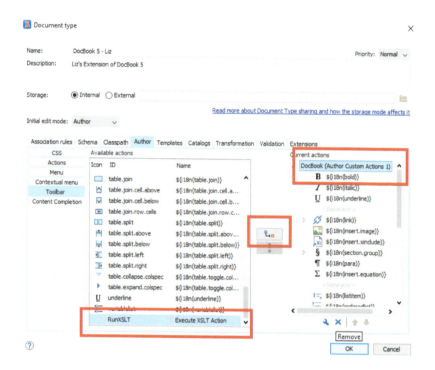

You will see the new button has been added to the toolbar currently visible in the **Current actions** pane.

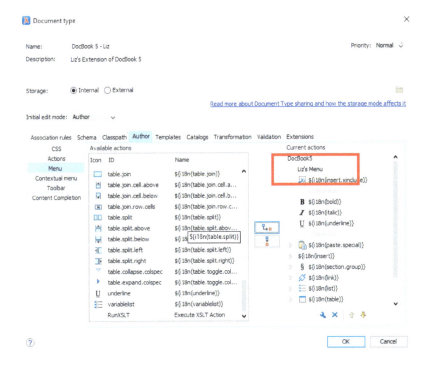

Click **OK** to return to the **Document Type** window.

Because you don't have an icon graphic defined, the button will display full text of the **Name** field you defined in your custom action. If you want to add an icon graphic to distinguish your custom action on the tool bar, you will need to edit the custom action and add the graphic icon to the definition.

Click **Actions** in the left-hand panel.

Adding Custom Actions to OxygenXML Frameworks

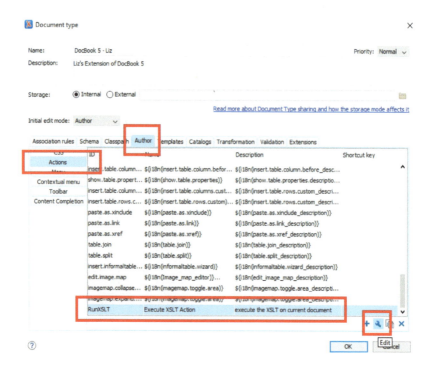

Select your custom action and click the **wrench** to edit it.

In the top half of the window, you can define the icon for this custom action. Click **Browse** and navigate to an icon graphic.

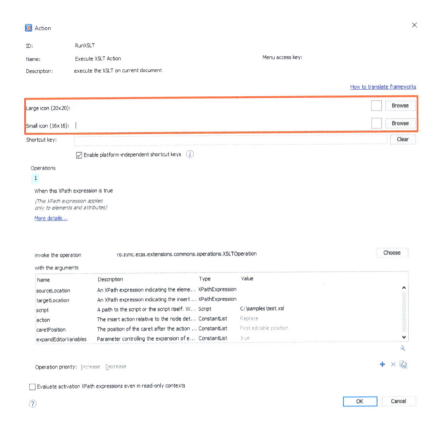

If you have a graphic that is usable, you will see it in the preview box.

Adding Custom Actions to OxygenXML Frameworks

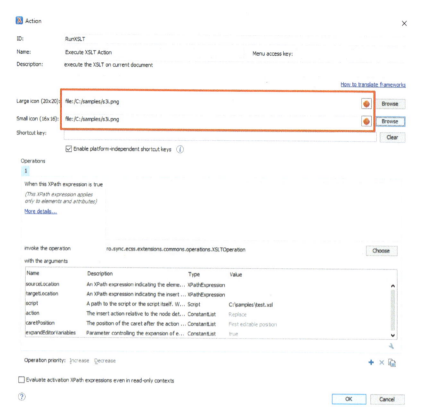

Click **OK** to save the custom action and to return to the Document Type window.

Click **Toolbar** in the left-hand pane and confirm that the icon now shows in the toolbar and action definition.

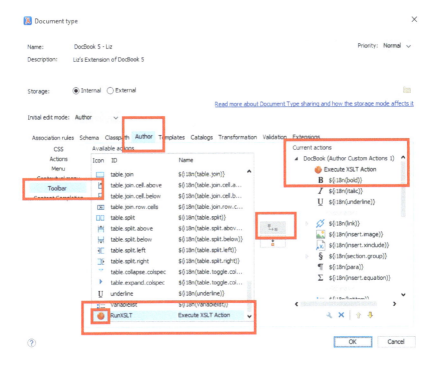

Execute your custom action

These are the basics for configuring custom actions and assigning them to menus and toolbar buttons. There are two things to remember about what you've done and how you did it.

First, because the custom action is attached to your framework extension, they will only appear if you are in a document that is based on your document type.

Adding Custom Actions to OxygenXML Frameworks

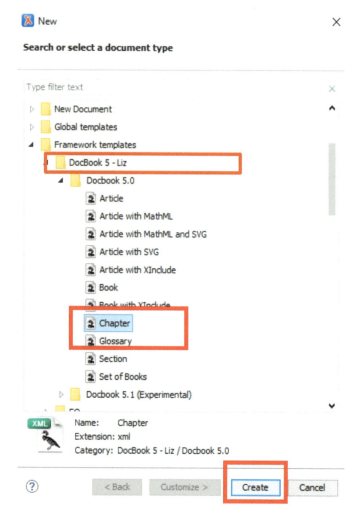

Second, because everything was configured in the Author tab, you will only see the custom menu and toolbar buttons when you are in author mode in OxygenXML.

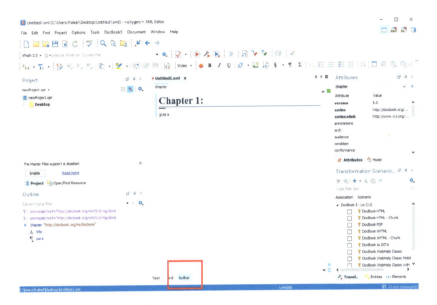

Test things out. Create a new document, from your extended doctype. Verify you are in author mode. You will see your custom action in your custom menu.

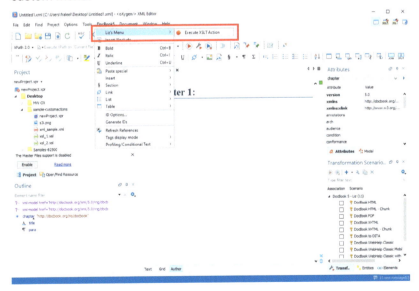

You will also see your icon on the toolbar.

Adding Custom Actions to OxygenXML Frameworks

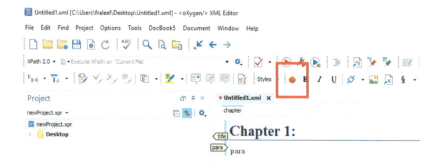

Both will execute your custom action!

You're ready to start creating custom actions of your own! Good luck!

Chapter 4.   Resources

Topics Covered in this Chapter

♦ Resources at OxygenXML.com
♦ Programming language references
♦ TC Dojo
♦ Mentoring @ Single-Sourcing Solutions

Although this is the first retail book about OxygenXML that is not part of the software user documentation or written by SyncRO Soft, many books use it in as the example software for teaching XML.

No book on a software technology would be complete without a list of resources available. I know things change, but at this time, all of these resources have been around more than 4 years (one more than 10!), so I feel confident including them here. Even older content often has value. When something has been around for mover 10 years, you take a lot of knowledge for granted. Things that were explained when the technology was new aren't explained 10 years later. User interfaces and features may have changed over time, but the underlying technology and information hasn't.

Here is the list of resources that can help you if you're starting to do any of the tasks described in this book:

Resources at OxygenXML.com

The OxygenXML website has a number of resources available — online documentation, support site, forums, videos, and github respository (DITA).

The staff monitors the forums and will respond to email. In fact, the team there is perhaps some of the best at responding to customers. They support a lot of different types of customers and go out of their way to be helpful. They are quick to understand what you need,

♦

providing exactly the right assistance, even when you aren't quite sure yourself.

Programming language references

If you want to learn more about XSLT and XQuery, we recommend the following:

- *Definitive XSLT and XPath* by G. Ken Holman. You can get it free from his website: cranesoftwrights.com

- *XQuery: Search Across a Variety of XML Data* by Priscilla Walmsley

Both books provide excellent resources for learning the intricacies of the languages in question.

Ken Holman has long been well known as the premier trainer for XSLT and XPath. If you learn better through video, he has videos of his training courses on Udemy: www.udemy.com/practical-transformation-using-xslt-and-xpath/.

TC Dojo

Location: www.tcdojo.org

The TC Dojo is a series of webinars in which the topics are chosen by the community and driven by community needs. Sessions are free to attend and recordings are published to the TC Dojo channel on YouTube (youtube.tcdojo.org).

The TC Dojo also has a Mastermind Group that is members-only. The group meets monthly and is focused on general technical communications issues. The group is part discussion group, part support group, part networking, and part training. It is governed by the members and has been described as,

"more useful than I ever could have imagined."

This group is not free. Members discuss real business issues and confidentiality is required. As a result, there is a small fee to attend to guarantee that random attendees are excluded. Members use a variety of tools and technologies and are not limited in their discussions or experiences.

Industry professionals have identified the TC Dojo as
"a technical communications resource that works."

Mentoring @ Single-Sourcing Solutions

Our customers want to do it themselves, but they don't want to do it alone. If you're reading this book then you, too, probably want to do it yourself.

As you develop your technical and conceptual skills, you might still have questions, or develop new ones. No book or video can cover every situation. You can always email us (info@single-sourcing.com) if you have general questions. Getting answers to questions that are specific-to you often require context and conversation. One way to facilitate that is to sign up for some kind of mentoring.

Think of going from journeyman to apprentice. You're working hard to grow your skills. You can't ask questions of a video or a book and sometimes you can't wait for days to get an answer from public forums. Our customers have found that a mentorship relationship boosts their confidence and helps them climb the learning curves faster.

Each of us at Single-Sourcing Solutions earned our expertise through carefully guided training that we received at the hands of mentors and talented coaches. Today, we are the mentors and coaches to our customers. Our mentoring services support the do-it-yourselfer as much as we support our full-service clientele.

Adding Custom Actions to OxygenXML Frameworks

We know that you want to take ownership of your project, your tools, and your content. Ultimately, you need to know how to maintain, grow, and extend them as your environment changes. When you work with a mentor, you don't simply outsource all that learning and responsibility. We help you to take ownership.

Chapter 5.   Afterword

Thanks for reading. I hope the book has been useful. I also did a webinar for the TC Dojo. The video has been posted to the website: oxygenxml.single-sourcing.com

Join the mailing list and you'll get more useful tips every month. We share information like candy! We pass more tips and resources along all the time. join.single-sourcing.com

Have Questions?

If you have questions about anything in the book or the technology that surrounds OxygenXML—XPath, XSLT, or XSL-FO—I can help. Need someone to help you write that XPath query or figure out why a template isn't being called? You can book an appointment with me directly! Get help for your stylesheet, customization, or any other question that applies specifically to you and your situation.

Never struggle for more than 30 minutes. Get help whenever you need it. You can book an appointment and get your questions answered:ask.single-sourcing.com

I'm here to help.

Tweet what you've read!

twitter.com/SingleSourcing

If Twitter's not your thing, everyone at Single-Sourcing Solutions participates in community projects—and we have a lot of them! To take advantage of one of our public service projects, go to social. single-sourcing.com and pick the one that works best for you.

Index

X